ON THE HUNT WITH LIONS

SANDRA MARKLE

Lerner Publications ◆ Minneapolis

THE ANIMAL WORLD IS FULL OF PREDATORS.

Predators are the hunters that find, catch, and eat other animals—their prey—in order to survive. Every environment has its chain of hunters. The smaller, slower, less able predators become part of the prey for the bigger, faster, more cunning hunters. And everywhere, there are just a few kinds of predators at the top of the food chain. These are the top predators. In Africa's grasslands, one of these is the lion.

Why are lions top predators? For one thing, the adults are big enough to hunt and kill their habitat's big prey, such as Cape buffalo. But they can't do that unless they live and hunt in family groups called prides. Overpowering a Cape buffalo takes at least five lionesses (females) or two lions (males). The males are bigger and more powerful. At 10 feet (3 m) long, including their tails, they're much longer than lionesses and a lot heavier. A full-grown lion may weigh as much as 550 pounds (249 kg), while a lioness might only weigh 395 pounds (179 kg). But both are strong.

Lions also use their size and strength to defend their pride's territory. That is an area of land where they can count on finding water and prey without traveling long distances. That's especially important for lionesses to successfully raise cubs—their babies.

A group of lions—two or three unrelated males or up to nine brothers and male cousins—is likely to team up. Together, they form a coalition to defend a prime territory. The pride leaders that make up the coalition patrol their territory's borders, roaring loudly to announce their claim. They also spray urine on rocks and bushes to create a scent fence. If they encounter any intruding lions, they attack to drive them away. But sometimes, an intruding coalition is stronger and forces out the current reigning males. Then the new lions take over the pride.

WOW!
A lion's roar can be heard up to 3 miles (5 km) away.

Another reason a lion is a top predator is its keen senses. A lion has an excellent sense of smell. The Jacobson's organ is a special body part in the roof of a lion's mouth that helps it sense smells.

To smell, a lion opens its mouth just a bit in what looks like a grimace. Then it sucks air in and over Jacobson's organ. This helps it find prey and locate kills made by other predators.

WOW!
For a short burst, a lion can run as fast as 50 miles (80 km) per hour.

A lion also has a keen sense of hearing. Its big ears on top of its head are sound scoops. But the sense a lion depends on most when hunting is its sight. Its big eyes give it a broad view of what's around it. A lion's forward-looking eyes let this big cat easily judge how far it has to run or leap to catch its prey.

A lion can also see to hunt in very dim light when it's easier to be sneaky. That's because its pupils—the dark spots in its eyes—can open extra wide to let light enter and strike the retina. This special layer at the back of each eye is covered with light-sensitive cells that send signals to the brain. The brain interprets these signals to allow the lion to see.

WOW!
A lion's reflective eye layer is called the tapetum lucidum. It causes a lion's eyes to glow when light strikes them at night.

A lion also sees well in dim light because its eyes have a mirrorlike layer behind the retina. This layer reflects light back through the retina and onto the light-sensitive cells, giving them a second chance to detect light signals. While this doesn't let a lion see in total darkness, it can see in much dimmer light than humans can. Keen eyesight in dim light helps a pride hunt and kill big prey, such as Cape buffalo, for a starlit dinner.

A lion's weapons, such as its claws, are another reason it is a top predator. Its claws let this big cat grab and hold onto prey, such as antelope. Though made of keratin like human fingernails, an adult lion's claws are stronger, up to 1.5 inches (3.8 cm) long, and very sharp.

To keep these weapons sharp, a lion's front paw claws are only exposed when needed. The big cat flexes or bends its toes to push the claws out of their protective sheaths. When it relaxes its toes, the claws go back inside their sheaths. Lions walk and run on their toes, so their claws would become dull by rubbing on the ground if they were not protected. Lions regularly scratch a tree trunk to file their claws and help keep them sharp.

A lion also has a mouthful of thirty strong, sharp teeth. And what weapons those are! A lion kills with a throat bite or by opening wide and clamping a bite over the prey's nose and mouth to stop its breathing.

A lion's front teeth are incisors, the smallest of its teeth, but just right for gripping and tearing. On either side of the incisors are the lion's daggerlike 3.9-inch-long (10 cm) canines. Behind the canines are flatter but still sharp premolars and special cutting teeth called carnassials. The carnassials are just right for slicing.

What really powers up an adult lion's bite are its jaws, which can open to 11 inches (28 cm) wide. That helps the lion slam its teeth down with a bite force that can crush bones!

WOW!

Though a lion's jaws are powerful, a lioness carries a cub in her mouth so gently she doesn't even prick its skin.

Another reason lions are top predators is the pride's teamwork. Though the males have the muscle power needed to bring down the biggest prey, they're slow. Their big showy manes also make them easy for prey to spot. So hunts that require speed or stealth are carried out by the lionesses.

The lionesses' tawny coats let them blend in with their surroundings to sneak up on prey, such as zebras. Once they get close, the females drop to their bellies and slip even closer until their prey senses the ambush and bolts.

The lionesses race in hot pursuit. Legs pumping, long tails flying to help them stay balanced, the big cats run flat out. The chase goes on until a weaker zebra lags behind the herd. Then the hunters focus on that prey.

Some lionesses hang back and let the first to attack do the work. Any attack requires energy, and lionesses risk getting injured as prey fights back. A predator is only after its next meal, while the prey is trying to save its life. That's why even top predators are often unsuccessful in their hunting efforts.

Sometimes two lionesses working together can overpower a zebra. One swipes the prey's back legs, causing it to stumble. Then both hunters pile on, dragging the zebra down. The kill is made when one lioness bites the prey's throat. She may still be gripping tight to make sure the zebra is dead when other pride members join in. The predators eat the soft body parts inside the zebra first because they are easily digested, high-energy food.

When the pride's males are nearby and hungry, they come to the kill. They take advantage of their size to eat first. The females usually back off and let them.

Lions don't chew their food. They bite off chunks or rip off mouthfuls with their scissorlike carnassial teeth. Then they gulp down the food, letting their digestive systems finish the job.

The most nutritious parts may be gone by the time the lionesses start eating. But they eat all they can to be strong for their next hunt. Any mothers with nursing cubs need plenty of food to be able to produce milk. The lionesses also allow cubs to grab bites of the kill. This way, the young lions begin eating solid food. Every meal gives the cubs a taste of what they will one day hunt.

WOW!

If enough food is available, each adult lion will eat as much as 40 percent of its body weight in a meal. That can be around 88 pounds (40 kg) of meat.

After a meal, the lions settle down to rest. They use their rough tongues to clean their coats—both their own and the coats of other pride members. No lion can lick its own head, so they help one another. Group grooming also strengthens the pride's sense of being a family. When their coats are clean, the big cats nap.

WOW!

Lions may spend eighteen to twenty hours a day sleeping.

Lions are top predators for one more reason. Lion prides share protecting, raising, and training their cubs—the next generation of hunters.

A lioness goes away from her pride to give birth. She usually has two or three cubs. She hides them in thick brush for about a month until their eyes are open and they can walk. During that time, she keeps track of her pride members by listening to them roaring. She only leaves her youngsters briefly each day to drink water and hunt prey she can quickly catch, such as a warthog or gazelle. If her pride roars to announce they made a kill nearby, she joins them for a meal.

Once the mother brings her cubs to the pride, the youngsters become part of a crèche, a group of cubs born around the same time. From then on, all the mothers share keeping an eye on the cubs to protect them.

When they aren't nursing and sleeping, the cubs sneak up and pounce on crèche mates. They also chase and wrestle one another. That's how they begin to develop their hunting skills. These games also help them become strong. Sometimes the lionesses hunt where the cubs have a clear view of the action. Those are the youngsters' earliest hunting lessons.

WOW!
Young lions make a lot of noises, but neither males nor females are able to roar until they are about two years old and their voice boxes have fully developed.

By the time cubs are three months old, they're doing a lot of exploring and learning to use their senses. They're also learning how to make the most of their growing size and strength. Sometimes they go along with the lionesses on hunts. The youngsters don't join in yet, but they pay attention to what happens. Since not every hunt delivers a meal, the cubs see what works and what doesn't.

When they're fourteen months old, the young adults are trying to catch prey on their own. They may even try chasing prey that's too big for them, such as a giraffe. While sure to fail, learning that hunting big prey takes the pride's teamwork is a valuable lesson. It's also a dangerous lesson since a giraffe's kick can kill.

WOW!
By the time a young male is ten months old, he's starting to grow his mane, though it's scruffy at first.

At about sixteen months, young adult lions start participating in hunts with the lionesses, but they still have a lot to learn. When sneaking up on prey, they're likely to charge before they're close enough to strike. This can mess up the hunt for everyone. But the lionesses don't punish the young hunters. Trying and failing is how the adults learned when they were training.

Every mistake that leads to a lost meal teaches a valuable lesson. And anytime the youngsters join in a successful hunt, full stomachs make the experience unforgettable.

A lion pride may be just a few big cats, or it may be more. A bigger group can claim better territory, and with each litter of cubs, a pride grows stronger. On average, a pride includes two or three adult males and five to ten adult females with their young. Females tend to stay with the same pride for life. But males, once they're about two to three years old, are pushed out by the resident males.

Young males may roam solo or team up with other males to form a coalition. By the time the males in a coalition are about four years old and at full strength, they are likely to be able to drive reigning males out of a pride. Then the coalition joins with that pride's lionesses.

For lions, being part of a pride means membership in a family team. It's how lions work together to defend their territory from other lion prides. It's also how lions join forces to hunt big prey, such as giraffes, so every pride member has plenty to eat. The lions work together to raise generation after generation of strong, successful hunters to maintain their pride as top predators!

A NOTE FROM SANDRA MARKLE

Even top predators face challenges, and global changes in Earth's climate are affecting lions. Longer dry periods mean Africa's grasslands aren't able to support as many large herds of grazing animals such as Cape buffalo, wildebeests, and zebras. These are the big prey animals that supply lions with the most food per kill. With these food sources shrinking, lion prides can't successfully raise large numbers of cubs to adulthood and become even larger family teams. The impacts of climate change stretch across the food chain in Africa's grasslands. Only time will tell how strongly these changes affect Africa's top predators, including lions.

Photo by Skip Jeffery Photography

LIONS SNAP FACTS

ADULT SIZE

Males are larger than females. Males may be up to 10 feet (3 m) long, including the tail, and weigh as much as 550 pounds (249 kg).

DIET

They mainly hunt and eat medium- to large-hoofed animals such as wildebeests, zebras, and gazelles.

RANGE

Lions mainly roam Africa's grasslands.

FUN FACT

The color of a male's mane isn't an inherited shade like human hair color is. Instead, a dark mane is a sign of the lion's strength and good health. No wonder scientific studies showed lionesses are attracted to lions with black manes!

LIFE SPAN

In the wild, lions may live up to nineteen years.

YOUNG

Females breed every two to three years. Cubs develop inside their mother for about 110 days. Then the female may only give birth to one cub or as many as six cubs.

GLOSSARY

CARNASSIAL: a type of tooth that is just right for slicing and cutting

COALITION: a group of male lions that hunt and travel together for their shared benefit, including taking over a pride

CRÈCHE: a group of cubs about the same age who stay together in a pride for one to two years and whose mothers share protecting, feeding, and training them

CUB: a young lion

FOOD CHAIN: a series of living things where each is dependent on another as a source of food

NURSE: to feed on milk from a mother's body

PREDATOR: an animal that hunts other animals

PREY: an animal that a predator catches to eat

PRIDE: a group of lions that live and hunt together

TAWNY: a warm, sandy color

INDEX

Image credits: Paul A. Souders/Getty Images, p. 3; AfriPics.com/Alamy Stock Photo, pp. 4, 5; Stef Bennett/Alamy Stock Photo, p. 6; Steve Bloom Images/Alamy Stock Photo, p. 7; Wim van den Heever/Getty Images, p. 8; WorldFoto/Alamy Stock Photo, p. 9; Avalon.red/Alamy Stock Photo, p. 11; Brigitta Moser/Getty Images, p. 12; infocusphotos.com/Alamy Stock Photo, p. 14; Moonstone Images/Getty Images, p. 15; saxreporter/Getty Images, p. 16; Michael Mantke/Alamy Stock Photo, p. 18; Stephen Frink/Getty Images, p. 19; Manoj Shah/Getty Images, p. 20; C.O. Mercial/Alamy Stock Photo, p. 21; Lori Ellis/Alamy Stock Photo, p. 22; AfriPics.com/Alamy Stock Photo, p. 23; Per-Gunnar Ostby/Getty Images, p. 25; Riaanvdb/Getty Images, p. 26; African Safari Collection/Alamy Stock Photo, p. 27. Design element: Exotic vector/Shutterstock.com.

Cover: Mike Hill/Alamy Stock Photo.

THE AUTHOR WOULD LIKE TO THANK DR. CRAIG PACKER, UNIVERSITY OF MINNESOTA LION CENTER, MINNEAPOLIS, MINNESOTA, FOR SHARING HIS ENTHUSIASM AND EXPERTISE.

A SPECIAL THANK-YOU TO SKIP JEFFERY FOR HIS LOVING SUPPORT DURING THE CREATIVE PROCESS.

FOR
RACHEL WEATHERFORD AND ALL THE CHILDREN AT BEAR CREEK ELEMENTARY SCHOOL IN HOUSTON, TEXAS

Lerner Publications Company
An imprint of Lerner Publishing Group, Inc.
241 First Avenue North
Minneapolis, MN 55401 USA

For reading levels and more information, look up this title at www.lernerbooks.com.

Main body text set in Aptifer Slab LT Pro medium.
Typeface provided by Linotype AG.

Library of Congress Cataloging-in-Publication Data

Names: Markle, Sandra, author.
Title: On the hunt with lions / Sandra Markle.
Description: Minneapolis, MN : Lerner Publications, [2023] | Series: Ultimate predators | Includes index. | Audience: Ages 8–12 | Audience: Grades 2–3 | Summary: "Lions are deadly predators. Their size, keen eyesight, and razor-sharp teeth and claws help them take down prey. Find out how lions hunt, interact with other lions in their pride, and raise their young"—Provided by publisher.
Identifiers: LCCN 2021056342 (print) | LCCN 2021056343 (ebook) | ISBN 9781728456294 (library binding) | ISBN 9781728464411 (paperback) | ISBN 9781728462448 (ebook)
Subjects: LCSH: Lion—Juvenile literature. | Predatory animals—Juvenile literature.
Classification: LCC QL737.C23 M2727 2023 (print) | LCC QL737.C23 (ebook) | DDC 599.757—dc23/eng/20211122

LC record available at https://lccn.loc.gov/2021056342
LC ebook record available at https://lccn.loc.gov/2021056343

Manufactured in the United States of America
1-50699-50118-3/22/2022